Solar Energy Systems Conference 2019

Waco, Texas, USA
12 – 13 December 2019

ISBN: 978-1-7138-0580-9

Printed from e-media with permission by:

Curran Associates, Inc.
57 Morehouse Lane
Red Hook, NY 12571

Some format issues inherent in the e-media version may also appear in this print version.

Printed with permission by Curran Associates, Inc. (2020)

For permission requests, please contact AIChE
at the address below.

AIChE
120 Wall Street, FL 23
New York, NY 10005-4020

Phone: (800) 242-4363
Fax: (203) 775-5177

www.aiche.org

Additional copies of this publication are available from:

Curran Associates, Inc.
57 Morehouse Lane
Red Hook, NY 12571 USA
Phone: 845-758-0400
Fax: 845-758-2633
Email: curran@proceedings.com
Web: www.proceedings.com

Solar Energy Systems Conference 2019

Waco, Texas, USA
12 – 13 December 2019

AIChE
The Global Home of Chemical Engineers

TABLE OF CONTENTS

SESSION 1 - SOLAR ENERGY STORAGE

SESSION 3 - MATERIALS FOR SOLAR THERMOCHEMISTRY AND PV

SESSION 4 - GRID INTEGRATION OF SOLAR ENERGY

POSTER SESSION

Invited Talk: Photocatalytic Particle-Suspension Reactors for Solar Water Splitting

Conference: Solar Energy Systems Conference

Year: 2019

Proceeding: Solar Energy Systems Conference '19

Group: Abstract Submissions

Session: Session 1 - Solar Energy Storage

Time: Thursday, December 12, 2019 - 10:30am-11:00am

Authors:

Bala Chandran, R., Lawrence Berkeley National Laboratory

Sunlight driven water splitting to produce hydrogen and oxygen provides a pathway to store available solar energy in the form of stable, energy-dense chemical bonds. Here, we investigate a tandem, particle-suspension reactor design for solar water splitting comprising micron-scale semiconductor particles (photocatalysts) suspended in an aqueous solution with soluble redox shuttles. Redox shuttles mediate electrons between the oxygen and hydrogen evolution reaction compartments. A porous separator facilitates redox species transport while averting gas crossover between the two reaction compartments. Numerical models have been developed to evaluate the transport of charged and neutral species, and reaction kinetics for the proposed reactor design at varying lengthscales ranging from that of a single photocatalyst particle to a suspension of photocatalyst particles in the reactor. At the lengthscale of a single particle, model results illustrate the impacts of particle size and the electrochemical potential of the redox shuttle on the rate of charge-carrier transport across the semiconductor-electrolyte interface. Using a transient reactor model, we demonstrate that a 1% solar-to-hydrogen conversion efficiency can be sustained by diffusion-driven mass transport in 1 cm tall reaction compartments using $BiVO_4$ and $SrTiO_3$:Rh photocatalysts for oxygen and hydrogen evolution, and with proton-coupled electron transfer redox shuttles. This talk will also highlight recent developments in my research group on designing and developing solar-powered wastewater treatment devices for energy and nutrient recovery.

Solar Energy, Opportunities and Challenges

Conference: Solar Energy Systems Conference

Year: 2019

Proceeding: Solar Energy Systems Conference '19

Group: Abstract Submissions

Session: Session 1 - Solar Energy Storage

Time: Thursday, December 12, 2019 - 11:30am-11:48am

Authors:

Ibeh, W. I. Sr., University

The fundamental economic reality of fossil fuels is that such fuels are found only in a relatively small number of locations across the globe, yet are consumed everywhere in the world. The economic reality, by contrast, is that solar resources are available, in varying degrees, all over the world. Fossil fuel and solar resource use are thus poles apart â€“ not just because of the environmental effects, but also because of the fundamentally different economical, logical and differing political, social and cultural consequences. These differences must be acknowledged if the full spectrum of opportunity for solar resources is to be exploited. Solar Energy a clean renewable, never-ending and sustainable resource with zero emission has got tremendous potential of energy which can be harnessed using a variety of devices. With recent developments, solar energy systems are easily available for industrial and domestic use with the added advantage of minimum maintenance. Solar energy is becoming more economically attractive as the government steps in, technologies improve, and the cost of electricity generated by fossil fuels rises. This development in solar energy is not without constraints as storage, location, space etc. are some of its constraints. These challenges can be curbed or minimized if recommended solutions to these challenges are implemented. The actions these players take will determine the solar sectorâ€™s scale, structure, and performance for years to come. This paper analyzes the innovation, opportunities and challenges of solar energy as an infinite source for power generation globally.

Keywords: Solar Energy, Photovoltaic, Challenges, Opportunities

Thermal Energy Storage Using the Boudouard Reaction

Conference: Solar Energy Systems Conference

Year: 2019

Proceeding: Solar Energy Systems Conference '19

Group: Abstract Submissions

Session: Session 1 - Solar Energy Storage

Time: Thursday, December 12, 2019 - 11:48am-12:06pm

Authors:

Miao, Y., Baylor University
Li, C., Baylor University
von Jouanne, A., School of Engineering and Computer Science, Baylor University
Yokochi, A., School of Engineering and Computer Science, Baylor University

This study presents a firm, indefinite storage of solar heat via chemical conversion of simple carbon-based chemicals using both the forward and reverse Boudouard reaction:

$C + CO_2$ â†” 2CO (Î”H = 171 kJ/mol)

This system is intended to be coupled with an external process like a supercritical CO_2 power turbine or a chemical conversion process. Depending on the operational needs, the energy storage reaction may be conducted to optimize performance at temperatures ranging between ca. 900K and 1100K (~600Â°C to ~800Â°C) by modifying the initial loading ratio of C and CO_2 into the system. In the thermal energy storage system, when excess heat is present as signaled by high temperatures in the external process, some of the heat is absorbed and stored as chemical energy, and when insufficient heat is supplied by the solar field, as reflected by lower temperatures in the external process, the heat is reinjected into it. In our design, the Boudouard reaction takes place in a fluidized bed reactor in which process solids reside, and where the heat is coupled into or out of the power cycle working fluid through the use of bidirectional heat pipes. The gas reactants/products are stored in packed beds of zeolitic material (copper doped zeolite 5A for CO, and simple ZSM-5 for CO_2). Systems at the 20 MW/320 MWh scale have been designed and modeled.

Current results reflecting system performance, modulating charge and discharge temperatures, and exergetic efficiency will be presented.

Optimal Sizing of Converters for Hybridizing Conventional Pumped Storage Hydro with Floating PV Plants

Conference: Solar Energy Systems Conference

Year: 2019

Proceeding: Solar Energy Systems Conference '19

Group: Abstract Submissions

Session: Session 1 - Solar Energy Storage

Time: Thursday, December 12, 2019 - 12:06pm-12:24pm

Authors:

Nag, S., Baylor University
von Jouanne, A., School of Engineering and Computer Science, Baylor University
Lee, K. Y., Baylor University
Yokochi, A., School of Engineering and Computer Science, Baylor University
Suchitra, D., SRM University

Pumped storage hydro (PSH) now has an expanded portfolio including adjustable speed (AS) synchronous machines to provide pump-mode regulation. However, the application of this technology in the industry has been limited to two plants world-wide, mainly due to the capital investment required for large-scale inverters. Also, the fact that an AS-PSH plant is idle during most of the day discourages the application of such investments.

To encourage investors and industrial participants to investigate this option, a multi-use case has been developed by coupling floating photovoltaics (PVs) and AS-PSH. The presentation proposes that the use of the existing but converters of the AS-PSH system for hosting floating PV will increase the availability of the plant and thereby justify the capital investment. Synchronous machine-based AS-PSH have large converters which can be used to integrate floating PVs of equal capacity as an AS-PSH unit, given the availability of the large amount of unused reservoir surface area. The large number of existing conventional PSH (C-PSH) projects world-wide presents significant opportunities for this hybrid technology to be realized.

Converters required for the conversion of the C-PSH plant to the AS-PSH and to the hybrid AS-PSH + PV plant can be optimally sized, as will be presented. The problem is framed as a security constrained optimal scheduling problem that simultaneously reveals the optimal size of the required converters. Benefits of this conversion are examined from the point of view of operating cost, emissions, losses, voltage profile, curtailment and system congestion under different levels of renewable penetration.

Toward Exploiting and Managing Solar Energy in Concentrated Solar Collectors: Using Models and Computational Strategies to Assess Solar Potential and Hourly Solar Irradiance Data Considering Sky Conditions

Conference: Solar Energy Systems Conference

Year: 2019

Proceeding: Solar Energy Systems Conference '19

Group: Abstract Submissions

Session: Session 1 - Solar Energy Storage

Time: Thursday, December 12, 2019 - 12:24pm-12:42pm

Authors:

Al-Aboosi, F. Y., Texas A&M Univeristy

The design and operation of concentrated solar thermal collectors, photovoltaic plant, energy storage unit are based on the precise assessment of solar potential and solar radiation data for a selected site. Particularly, direct normal irradiance that used in the long-term evaluation for the techno-economic performance of solar energy conversion systems. In addition to the high cost of instruments used in measuring stations, there is a significant lack of direct normal irradiance data comparing to global and diffuse horizontal irradiance data. To find the proper methods that can contribute to terminating challenges of creating solar databases, this study presented two hierarchical methodologies using different models, empirical correlations, and regression equations. These models and methodologies can be used in estimating hourly solar irradiance data for a selected location using new correlation coefficients and various sky conditions using cloud cover range. Furthermore, the preliminary decision to determine the ability to exploit solar energy technologies was achieved through developing a comprehensive analysis for assessing the potential of solar energy using the solar irradiance data and clearness index. The San Antonio region in Texas was chosen as a paradigm to elucidate the precision of the proposed calculation strategies for estimating hourly direct normal irradiance and monthly average hourly direct normal irradiance data. The obtained results show a significant potential of harnessing solar energy as one of the main energy sources at the selected region and good accuracy of estimated solar data comparing with measured solar data using locally adjusted coefficients and different statistical indicators.

Coordinate Control Strategy for Multiple Distributed Generations with PV in a Microgrid

Conference: Solar Energy Systems Conference

Year: 2019

Proceeding: Solar Energy Systems Conference '19

Group: Abstract Submissions

Session: Session 3 - Materials for Solar Thermochemistry and PV

Time: Friday, December 13, 2019 - 10:20am-10:38am

Authors:

Lee, K. Y., Baylor University

A coordinated control strategy will be presented to realize efficient power management in a microgrid which is connected to 12-node distribution system and consists of photovoltaic (PV) array, solid-oxide fuel cell (SOFC) and an ultra-capacitor (UC). The output power of a PV array is intermittent due to the variations of irradiation and temperature, causing a power imbalance in the power system, while SOFC is not only able to generate power to its own local load, but also to compensate the power imbalance caused by the intermittency of PV array. A power management is implemented by fuzzy logic control to coordinate the operations of SOFC and UC, maximizing the lifetime of SOFC and improving the transient response by ultra-capacitor as supplementary energy source. Therefore, the fuel utilization rate should be in the safe operation range, and the state-of-charge of UC should be maintained in the normal range to enhance the lifetime of UC. The proposed operating strategy always operates the PV array at the maximum power point and the SOFC in high efficiency and safe performance range, hence, to enhance the system stability and improve the performance of system operation.

Invited Talk: Scalable Nanomanufacturing of Bio-Inspired Optical Thin Films for Solar Photovoltaics

Conference: Solar Energy Systems Conference

Year: 2019

Proceeding: Solar Energy Systems Conference '19

Group: Abstract Submissions

Session: Session 3 - Materials for Solar Thermochemistry and PV

Time: Friday, December 13, 2019 - 10:38am-11:08am

Authors:

Chang, C. H., Oregon State University

microreactors for the synthesis, assembly, and deposition of nanomaterials. In synthesis, microreactor technology offers large surface-area-to-volume ratios within microchannel structures to accelerate heat and mass transport. This accelerated transport allows for rapid changes in reaction temperatures and concentrations, leading to more uniform heating and mixing in the deposition process. The possibility of synthesizing nanomaterials in the required volumes at the point-of-application eliminates the need to store and transport potentially hazardous materials while providing new opportunities for tailoring novel nanostructures and nanoshaped features. MAND processes control the heat transfer, mass transfer, and reaction kinetics using well-defined microstructures of the active unit reactor cell that can be replicated to produce higher chemical production volumes. This critical feature opens a promising avenue in developing scalable nanomanufacturing. Furthermore, the continuous flow microreactor opens up the opportunity to conveniently assemble unique nanostructures and nanostructured thin films. Results-to-date demonstrates the possibility to control the reacting flux, including small intermediate-reaction molecules, nanoclusters, nanoparticles, and structured assembly of nanomaterials.

In this presentation, recent progress in using continuous microreactors to synthesize bio-inspired nanostructured thin films and the applications of these functional nanostructured thin films for solar photovoltaics will be discussed.

Invited Talk: Cerium-Vanadium Metal Oxides for Syngas Production Via Solar Thermochemical Redox Cycles

Conference: Solar Energy Systems Conference

Year: 2019

Proceeding: Solar Energy Systems Conference '19

Group: Abstract Submissions

Session: Session 3 - Materials for Solar Thermochemistry and PV

Time: Friday, December 13, 2019 - 11:08am-11:38am

Authors:

Lipinski, W., The Australian National University
Riaz, A., The Australian National University
Tsuzuki, T., The Australian National University
Lowe, A., The Australian National University

Developing durable redox materials with fast and stable redox kinetics under high-temperature operating conditions is a key challenge for an efficient industrial-scale production of synthesis gas via two-step solar thermochemical redox cycles. CeO_2 possesses a high phase stability with moderately high and stable redox kinetics during the reduction and oxidation reactions. Materials with a high degree of catalytic activity improve the syngas yield. V_2O_5 is well-known for its catalytic activity in redox processes such as electrolysis, gas sensing and hydrocarbon oxidation. Here, the effects of V and Ce concentrations (each varying in the 0â€“100% range) in vanadiaâ€“ceria multi-phase systems are investigated for synthesis gas production via thermochemical redox cycles of CO_2 and H_2O splitting coupled to methane partial oxidation reactions. Stable redox rates are observed in pure CeO_2, while pure V_2O_5 is characterized by the highest syngas yield at the cost of irregular kinetic rates between cycles. Incorporation of vanadium in CeO_2 improves the syngas yield without impacting the stability of redox rates. The addition of 25% V to CeO_2 results in an optimum mixture of CeO_2 and $CeVO_4$ for enhanced CO_2 and H_2O splitting. At higher V concentrations, cyclic carbide formation and oxidation result in a syngas yield higher than that for pure CeO_2. The structural and chemical analysis of the mixed metal oxides provides insights into the phase transformation in V_2O_5 due to cerium and vanadium interactions.

Solar Thermochemical Energy Storage Using Blended Peroxides

Conference: Solar Energy Systems Conference

Year: 2019

Proceeding: Solar Energy Systems Conference '19

Group: Abstract Submissions

Session: Session 3 - Materials for Solar Thermochemistry and PV

Time: Friday, December 13, 2019 - 11:38am-11:56am

Authors:

Lei, F., Oregon State University
Dyall, A., Oregon State University
AuYeung, N., Oregon State University

Group II peroxide materials were investigated for solar thermochemical energy storage. Strategies were proposed to improve the longevity and reactivity. The proposed material matrix was able to operate longer cycles without deactivation, demonstrating the suitability of such system for solar thermochemical energy storage. Different techniques such as thermogravimetric analysis (TGA) and differential scanning calorimetry (DSC), XRD as well as SEM/EDS were used to study the material properties and performance.

Material Degradation Mechanisms Relevant to the Next Generation of Concentrated Solar Power Plants

SOLAR ENERGY SYSTEMS CONFERENCE

Conference: Solar Energy Systems Conference

Year: 2019

Proceeding: Solar Energy Systems Conference '19

Group: Abstract Submissions

Session: Session 3 - Materials for Solar Thermochemistry and PV

Time: Friday, December 13, 2019 - 11:56am-12:14pm

Authors:

Will, G., Queens
Steinberg, T., Queensland University of Technology

Abstract

The higher working temperature of future Concentrated Solar Power (CSP) plants requires careful selection of containment materials (high-strength alloys) to meet the challenging high-temperature and corrosive environments within the plant. This is also true of high-temperature thermochemistry derived from renewables. For CSP, the receiver will operate at ~800 °C with liquid salt or sodium, all fluid piping systems allowing the transport of the heat transfer fluids, heat exchangers that permit the heat to flow through the CSP plant, storage tanks (with phase-change materials) operating in excess of 700 °C storing the thermal energy collected for later use, and the supercritical CO_2 power block operating in excess of 600 °C are critical components that need to survive harsh, high-temperature, pressurized environments that have thermal cycles designed into their ongoing operation. Thermo-mechanical fatigue, transient creep, (cyclic) corrosion, carburization, embrittlement and erosion have been recognized as general degradation mechanisms for all containment (metallic material) sections of the system as well as degradation of the heat transfer and storage fluids (sodium, phase change materials, supercritical CO_2). Accordingly, and for the purpose of improving compatibility, characterising the various and significant degradation mechanisms for a matrix of candidate materials has been developed and is currently being evaluated through specific and targeted testing to validate their appropriateness for these future and challenging CSP applications. Candidate metallic materials selected, the testing being conducted and preliminary results for the degradation of these materials in the various heat-transfer fluids will be presented and a summary of ongoing/future work.

Stabilized Mo As a MOC for CSP and Solar Fuels

Conference: Solar Energy Systems Conference

Year: 2019

Proceeding: Solar Energy Systems Conference '19

Group: Abstract Submissions

Session: Session 3 - Materials for Solar Thermochemistry and PV

Time: Friday, December 13, 2019 - 12:15pm-12:30pm

Authors:

Trainham, J., University of Florida
Scheffe, J., University of Florida
Cook, B., RTI International

Solar fuels and chemical production and future concentrated solar power (CSP) all require high operating temperatures above 800 °C and some chemistries up to 1500 °C. Examples include water splitting to produce H_2 (~ 1400 °C), steam gasification of biomass or municipal waste (1200 °C) and methane pyrolysis to acetylene (1500 °C) and natural gas reforming (~ 1400 °C). Major barriers to achieving commercial success are materials of construction (MOC) and improving the chemistries to achieve higher yields and at lower temperatures.

To reduce the cost of CSP and solar-thermal chemistries, it is suggested that Mo stabilized in air with an MoSiB composite coating or the MoSiB composite itself can offer many advantages as

MOC. Mo is compatible with most molten metals and molten salts, has a tensile strength > 120 MPa at 1000 °C that will allow thin wall tubing and components. Results show stability in air to 1650 °C and thermal cycling 500 times to 1500 °C. The MoSiB composite was produced by a scalable, low cost mechanical alloying technique.

One application for this MOC is solar driven thermochemistry which takes place in the temperature range of 1000 â€“ 1400 Â°C. In this presentation, we will also discuss a scalable prototype receiver, which contains an indirectly-irradiated packed bed reactor, which was developed to experimentally demonstrate the chemical-looping, dry reforming of methane over ceria with simulated concentrated solar radiation

Dopant Effects on Ceria Solar Thermochemical Water Splitting Behavior

Conference: Solar Energy Systems Conference

Year: 2019

Proceeding: Solar Energy Systems Conference '19

Group: Abstract Submissions

Session: Session 3 - Materials for Solar Thermochemistry and PV

Time: Friday, December 13, 2019 - 12:32pm-12:50pm

Authors:

Muhich, C. L., University of Colorado at Boulder

Solar thermochemical water and carbon dioxide splitting provide a pathway to high efficiency solar fuels. To-date, ceria remains the benchmark material due to its stability and rapid kinetics. However, the low redox capacity of ceria has driven intense research into improving behavior. Here, we delve into the effects of dopant addition on the fundamental thermodynamic and kinetics of the water splitting process. We use quantum mechanical simulations to determine not only the effects of dopant addition, but the underlying cause. Though these models, we can see why some dopants are good, e.g. Zr, and why others are poor performers, e.g. Gd. This then can point towards novel materials design, particularly paired charge compensating dopants (PCCD). Additionally, experiment is used to test these new materials and validate computational results.

Invited Talk: Grid Service Provision with Inverter-Based Renewable Resources

Conference: Solar Energy Systems Conference

Year: 2019

Proceeding: Solar Energy Systems Conference '19

Group: Abstract Submissions

Session: Session 4 - Grid Integration of Solar Energy

Time: Friday, December 13, 2019 - 1:30pm-2:00pm

Authors:

Gravagne, I., Baylor University
Guttromson, R., Sandia National Labs

Inverter-based renewable resources have the potential to provide secondary grid services such as frequency and voltage support. Variations in transmission frequency or voltage within certain frequency bands, for example 0.1 to 2 Hz, can be suppressed through inverter power modulation assuming low-latency real-time feedback is available. First, we present experimental results that illustrate this idea using synchrophasor-based feedback to modulate a Vestas wind turbine in Lubbock, TX. Results here show that inter-area frequency oscillations can be counteracted with no net loss of steady-state turbine output by adding and subtracting from the turbineâ€™s spinning potential energy. Next we explore how similar inverter modulation services could be provided by large solar installations. We examine some of the key differences and challenges to solar-based services compared to wind-based, including design issues for grid instabilities over larger bandwidths and the tradeoffs between designs that utilize a small amount of storage vs operating the solar installation at less than 100% steady-state capacity. We also note that solar installations can in fact be a *cause* of grid instabilities, forcing operators to curtail. Inverter-based feedback stabilization may therefore enable higher overall output over existing transmission infrastructure.

Invited Talk: IEEE Standard Development for Inverter Based Resources Connected to the Transmission Grid

Conference: Solar Energy Systems Conference

Year: 2019

Proceeding: Solar Energy Systems Conference '19

Group: Abstract Submissions

Session: Session 4 - Grid Integration of Solar Energy

Time: Friday, December 13, 2019 - 2:00pm-2:30pm

Authors:

Guttromson, R., Sandia National Labs

This presentation will discuss the purpose and progress of the new IEEE Standard P2800 under development, Interconnection and Interoperability of Inverter-Based Resources Interconnecting with Associated Transmission Electric Power Systems. The global increase in penetration levels of inverter-based resources is expected to significantly change the dynamic performance of the power grid. As the penetration levels of inverter-based resources increase and this technology evolves, specifications and standards are needed to address the performance requirements of these resources. Currently, there is no one single document of consensus performance requirements covering inverter-based resources interconnected with transmission and sub-transmission systems. P2800 will ensure that related standards (such as P-1547) are not misused for transmission applications, and will help equipment manufacturers, project developers, transmission planners, and power grid operators improve the quality of the inverter performance to enhance the reliability and stability of the power grid.

Mapping the Future of Agrivoltaics

Conference: Solar Energy Systems Conference

Year: 2019

Proceeding: Solar Energy Systems Conference '19

Group: Abstract Submissions

Session: Session 4 - Grid Integration of Solar Energy

Time: Friday, December 13, 2019 - 2:30pm-2:48pm

Authors:

Brickhouse, M., Saint Louis University

Agrivoltaics is at the heart of food-water-energy nexus, a new way of combining renewable energy with agriculture to address a range issues including urban heat island effects, food deserts in urban environment while improving food security under climate change challenges.

Recent demonstrations have shown that the combination of photovoltaics and agriculture on shared land increases overall land productivity. It has also been observed than when used with shade-tolerant crops, PV installations reduce transpiration and improve water retention. However, research is needed to determine the geographical sites in which agrivoltaics can have the greatest impact on bringing marginal lands into agricultural production, and new areas opened to the production of high value crops through its combination of shading and improved water retention. We propose the analysis of currently available high-definition satellite imagery and hyper-spectral imagery of potential rural and urban agrivoltaics sites to determine current status of arability, and to use modeling based on knowledge of insolation and water loss reduction by photovoltaics (PV) to determine optimum siting and crop selection. The objectives of this research will include:

1. Improved Smart city and urban agriculture.
2. Determination of currently marginal lands which can be brought into production by agrivoltaics.
3. Optimization of shared light resources for energy and crop production.
4. Plant breeding advancement for shade tolerant varieties of crops and vegetables.

Using Artificial Intelligence to Accurately Forecast Solar Energy Including Air Quality in the Models

Conference: Solar Energy Systems Conference

Year: 2019

Proceeding: Solar Energy Systems Conference '19

Group: Abstract Submissions

Session: Session 4 - Grid Integration of Solar Energy

Time: Friday, December 13, 2019 - 2:48pm-3:06pm

Authors:

Rausch, N., SAS Institute

Accurate power forecasting is an important factor in photovoltaic (PV) solar array systems, due to the volatile nature of the technology. This paper proposes an advanced statistical method for accurate solar power forecasting 12-24 hours in advance using artificial intelligence techniques. The method incorporates previous power output from a solar panel array, local weather data, sun position, and irradiance levels as important inputs. In addition, a unique feature of this paper is the incorporation and study of the impact that local air pollution levels have on the forecast model. The results show that the Random Forest method produces a high-quality model with a very low error rate for medium term forecasts of 12-24 hours into advance, compared to other, more complex models such as a recurrent neural network. The study also highlights the influence that air quality pollution levels have as an input predictor to the model. The influence of air quality pollution is greatest from particle pollutants during winter months. The model proposed in this paper can be used as an efficient and easily interpretable forecasting technique for operational planning for a solar array.

Invited Talk: Real-World Observations of Grid Oscillations Due to Photovoltaics Via Synchrophasors

Conference: Solar Energy Systems Conference

Year: 2019

Proceeding: Solar Energy Systems Conference '19

Group: Abstract Submissions

Session: Session 4 - Grid Integration of Solar Energy

Time: Friday, December 13, 2019 - 3:06pm-3:36pm

Authors:

Mattei, A., Baylor University
Grady, M., Baylor University

The Texas Synchrophasor Network has been monitoring the ERCOT grid since 2009. During this time, PV installations within ERCOT have grown from 15 MW in 2009 to nearly 1900 MW in 2018, with significantly more solar capacity on the way. Frequent grid oscillations in the 1.4 to 10 Hz range have been captured by the phasor measurement units and logged by the synchrophasor data collection system. These oscillations correspond to daylight hours and seasonal variations in load. This article describes the Texas Synchrophasor Network, the procedure used for detecting the low-frequency grid oscillations associated with photovoltaics, and presents actual ERCOT grid oscillation data from multiple seasons.

Integrating Solar Energy into Micro-Grids for Smart Manufacturing Plants

Conference: Solar Energy Systems Conference

Year: 2019

Proceeding: Solar Energy Systems Conference '19

Group: Abstract Submissions

Session: Session 4 - Grid Integration of Solar Energy

Time: Friday, December 13, 2019 - 3:36pm-3:54pm

Authors:

Quazi, H., AngelNet Energy, Water and Environmental Systems Inc

MARTECH International Inc. and AngelNet Energy, Water & Environmental Systems Corporation have jointly developed an innovative solution for the smart manufacturing industries requiring additional, reliable power for their expanded operation. The solution "AngelNet-Microgryd" is a comprehensive optimized control system for the design and operation of Micro-Grids for reliable and affordable electricity. It integrates solar energy with conventional base-load sources and energy storage that eliminate the intermittency issue of solar generation. The technology contribute to a reduction in harmful emissions and enable providing reliable electricity to industries at competitive pricing.

Solar Farm Power Forecasting Using Deep Learning

Conference: Solar Energy Systems Conference

Year: 2019

Proceeding: Solar Energy Systems Conference '19

Group: Abstract Submissions

Session: Session 4 - Grid Integration of Solar Energy

Time: Friday, December 13, 2019 - 3:54pm-4:12pm

Authors:

de Castro, A., SAS Institute Inc.
Lee, T. Y., SAS Institute Inc.
Hwang, Y. T., Georgia State University
Lee, H. Y., Tmoney Co. Ltd.

Solar energy is the third largest renewable energy source in the world after hydro and wind power. Its role is expected to become even more significant as technology allows the production of cheaper and higher efficiency units, resulting in increased adoption and tremendous impacts on grid operations. Forecasting solar energy is therefore becoming more and more important.

Deep Learning models using RNN, LSTM, GRU are gaining popularity for electrical Load forecasting. The accuracy obtained with these techniques vary depending on the forecasting horizon. Different Deep Learning models work for Very Short Term (VST), Short Term (ST), and Long Term (LT) forecasting. In this research, we test various Deep Learning models at different time steps for solar farm power forecasting for each of these time frames, comparing with Machine Learning models of Random Forest, Gradient Boosting and Neural Networks.

Connecting of 540kW PV Solar to Grid

SOLAR ENERGY SYSTEMS CONFERENCE

Conference: Solar Energy Systems Conference

Year: 2019

Proceeding: Solar Energy Systems Conference '19

Group: Abstract Submissions

Session: Session 4 - Grid Integration of Solar Energy

Time: Friday, December 13, 2019 - 4:12pm-4:30pm

Authors:

Al Hadi, H., MOG
Baram, A., MOG
Al Rishadi, A., MOG
Al Ghafri, A., MOG

The PV solar System become one of the Oman national strategy as 2030, will be 30% as renewable , this paper will present the connection of 540kW of PV Solar to the grid at 415V distribution system , which was planned to reduce the consumption bill of the Ministry of Oil and Gas (MOG) , and to improve the power quality to the connected load. The paper will provide techno economical studies and will be presented through a power simulator. The paper will provide the challenges and opportunities for connecting this system to the grid, and highlights the current situation of Oman renewable strategy and requirements.

Detailed Balance Analysis of Photovoltaic Windows

Conference: Solar Energy Systems Conference

Year: 2019

Proceeding: Solar Energy Systems Conference '19

Group: Abstract Submissions

Session: Session 4 - Grid Integration of Solar Energy

Time: Friday, December 13, 2019 - 4:30pm-4:48pm

Authors:

Wheeler, V., The Australian National University
Wheeler, L., National Renewable Energy Laboratory

There are several technical and socio-economic factors converging to position photovoltaic (PV) windows as a transformative technology for the energy landscape of the future: 1) Urban areas currently account for 67â€“76% of global final energy consumption. 2) 70% of the worldâ€™s population will live in urban areas by 2050. 3) The overwhelming architectural trend is away from opaque building components and toward all-glass facades. 4) Photovoltaics are becoming extremely affordable, and the most expensive components in a conventional module are the glass and transparent metalsâ€”components that are already in highly-insulating glazing. Rational design of PV windows is of paramount importance to realize their impact. In this work, we provide an analysis on the theoretical performance of PV windows using a detailed balance model to understand the complex design space of power conversion efficiency (PCE), visible light transmittance (VLT), solar heat gain coefficient, and color. We find there are two distinct regimes for PV window absorber design. The first low-VLT regime validates the most prevalent approach to semitransparent PV windows in which conventional absorber materials are thinned to allow visible light to pass throughâ€"the bandgap for peak PCE is consistent with the classic work of Shockley and Queisser. However, we identify a second, high-VLT regime in which the ideal bandgap for maximum PCE increases monotonically from 2 to 3 eV with increasing VLT. In this tuned-bandgap regime, the solar cell exhibits lower losses and tunable solar heat gain and color.

Comparative Assessment of Solar Thermal Water Splitting Materials

Conference: Solar Energy Systems Conference

Year: 2019

Proceeding: Solar Energy Systems Conference '19

Group: Poster Submissions

Session: Poster Session

Authors:

Tran, J. T., Oregon State University
Weimer, A. W., University of Colorado Boulder
Warren, K. J., University of Colorado Boulder

Hydrogen is a clean energy source that can replace fossil fuels in the transportation and power sectors, two of the largest contributors of anthropogenic greenhouse gas emissions. Globally, more than 95% of hydrogen is produced from fossil fuels, resulting in CO_2 emissions and a significant carbon footprint. To address this, researchers have sought to produce hydrogen via solar thermal water splitting (STWS) using solar energy to drive the endothermic water splitting reaction via a reduction/oxidation cycle. Cerium (IV) oxide, or ceria, CeO_2 has demonstrated benchmark performance for STWS because it exhibits excellent thermal stability and fast redox kinetics. However, meaningful hydrogen productivities are only observed following reduction at temperatures above 1500°C, which are too high for commercial realization. Alternatively, perovskites have demonstrated water splitting at more suitable temperatures but have shown either lower rates or productivity as compared to ceria.

This study looks to compare different STWS materials and evaluate their performance under various operating conditions. Ceria and various perovskites (SLMA $Sr_xLa_{1-x}Mn_yAl_{1-y}O_3$, BCM $BaCe_{0.25}Mn_{0.75}O_3$, CSM $Ce_xSr_{2-x}MnO_4$, LSCM $La_{0.6}Sr_{0.4}Cr_xMn_{1-x}O_3$) were tested in a stagnation flow reactor. Materials are compared by evaluating their hydrogen production capacity and water consumed to hydrogen produced (H_2O:H_2) ratio. Results show that ceria and LSCM have low productivities, but fast kinetics while SLMA, BCM, CSM produce more hydrogen but over longer oxidation times for the conditions tested, showing a tradeoff between productivity and kinetics. A direct comparison provides researchers with a path forward to ensure that STWS is competitive with other hydrogen production technologies.

Heliostat and Solar Simulator for Solar Thermal Applications

Conference: Solar Energy Systems Conference

Year: 2019

Proceeding: Solar Energy Systems Conference '19

Group: Poster Submissions

Session: Poster Session

Authors:

Miao, Y., Baylor University
Li, C., Baylor University
Acton, S., Baylor University
Parker, L., Baylor University
Agwamba, C., Baylor University
Heidemann, A., Baylor University
Nahrgang, C., Baylor University
Collin, R., School of Engineering and Computer Science, Baylor University
von Jouanne, A., School of Engineering and Computer Science, Baylor University
Yokochi, A., School of Engineering and Computer Science, Baylor University

As a renewable power source, solar thermal energy is considered clean and inexpensive and can be harnessed nearly everywhere in the world. It represents a limitless source of power since solar energy comes from the sun. Unlike photovoltaic systems that directly convert sunlight into electrical energy during daylight hours, solar thermal technology can store the solar energy as heat and convert that thermal energy to electrical energy even after sunset or when the sun is blocked by clouds [1]. Other applications of solar thermal energy include chemical conversions (i.e., solar fuels and chemicals) and thermal separations (i.e., direct injection of solar heat into distillation towers).

To enable concentrating solar energy systems research in our labs, we have built a heliostat and a solar simulator. The design of the systems and characterization regarding irradiance level and uniformity at the receiver, light collimation, and spectral distribution will be presented.

Reference:

[1] Stutz, B., *et al.*(2017). â€œStorage of thermal solar energyâ€ Comptes Rendus Physique, 18: 401-414.

AUTHOR INDEX

AIChE
120 Wall Street, FL 23
New York, NY 10005-4020

ISBN 978-1-7138-0580-9

CURRAN ASSOCIATES INC.
proceedings.com